PLANTS with SUPERPOWERS

# PLANTS CAN EAT MEAT!

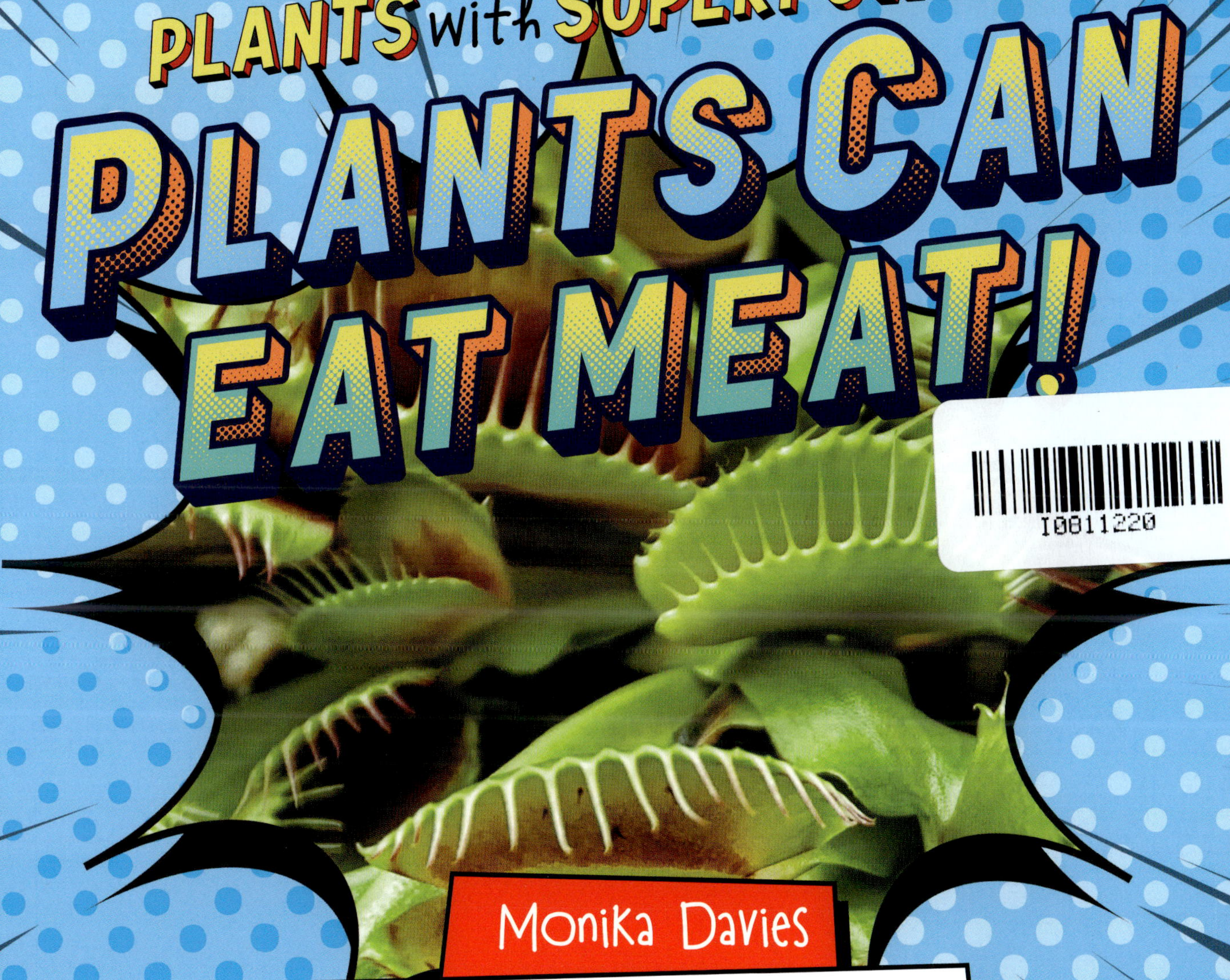



Monika Davies

Gareth Stevens PUBLISHING

Please visit our website, www.garethstevens.com. For a free color catalog of all our high-quality books, call toll free 1-800-542-2595 or fax 1-877-542-2596.

**Library of Congress Cataloging-in-Publication Data**
Names: Davies, Monika, author.
Title: Plants can eat meat! / Monika Davies.
Description: New York : Gareth Stevens, [2023] | Series: Plants with superpowers | Includes bibliographical references and index.
Identifiers: LCCN 2022006641 | ISBN 9781538278888 (set) | ISBN 9781538278895 (library binding) | ISBN 9781538278871 (paperback) | ISBN 9781538278901 (ebook)
Subjects: LCSH: Carnivorous plants–Juvenile literature.
Classification: LCC QK917 .D38 2023 | DDC 583/.887–dc23/eng/20220317
LC record available at https://lccn.loc.gov/2022006641

Published in 2023 by
**Gareth Stevens Publishing**
29 East 21st Street
New York, NY 10010



Editor: Monika Davies
Designer: Leslie Taylor

Portions of this work were originally authored by Keisha Jones and published as *Plants that Eat.* All new material in this edition authored by Monika Davies.

Photo credits: Cover, p. 1 Barbara Ash/Shutterstock.com; Series Art (various accents) Ozz Design, Piotr Urakau, Designer things, Maxger/Shutterstock.com; p. 5 Kuttelvaserova Stuchelova/Shutterstock.com; p. 7 (pitcher) JTKP/Shutterstock.com; p. 7 (sundew) John Ceulemans/Shutterstock.com; p. 9 (brown cobra) Mike Russell/Shutterstock.com; p. 9 (cobra) https://commons.wikimedia.org/wiki/File:My-13-4030-Cobra-Cobra-Lily-Cameron-Cobra-Lily-HiR.jpg; p. 11 https://commons.wikimedia.org/wiki/File:The_Captured_(15232198306).jpg; p. 13 (butterwort) blickwinkel/Jagel/Shutterstock.com; p. 13 (sundew) https://commons.wikimedia.org/wiki/File:Another_Sundew.jpg; p. 13 (pitcher) Jeff McGraw/Shutterstock.com; p. 15 (moth in trap) Carolina K. Smith MD/Shutterstock.com; p. 15 (grasshopper in trap) AlessandroZocc/Shutterstock.com; p. 17 (pitchers) jaboo2foto/Shutterstock.com; p. 17 (spiders) Linas T/Shutterstock.com; p. 17 (monkey cup) marchkimoo/Shutterstock.com; p. 17 (tropical pitcher) Photogrape/Shutterstock.com; p. 19 (map) https://commons.wikimedia.org/wiki/File:Utricularia_distribution.svg; p. 19 (trapped bladders) blickwinkel/Hartl/Alamy.com; p. 19 (yellow) Luka Hercigonja/Shutterstock.com; p. 19 (purple) anjahennern/Shutterstock.com; p. 21 (damselfly) Pim Leijen/Shutterstock.com; p. 21 (fly) Ernie Cooper/Shutterstock.com; p. 21 (sundew) https://commons.wikimedia.org/wiki/File:Sundew_fly.jpg.



Printed in the United States of America



CPSIA compliance information: Batch #CSGS23: For further information, contact Gareth Stevens, New York, New York, at 1-800-542-2595.

# CONTENTS



Words in the glossary appear in **bold** type the first time they are used in the text.

# Killer Plants

From far away, most plants seem harmless. Many grow bright flowers that add life and color to our gardens. Plants often have a soothing, or comforting, effect on us. But did you know some plants have a killer **instinct**?

Meet carnivorous plants, or meat-eating plants that grow wild across the world. These plants have bugs and animals on their dinner menu. They use traps to capture their next meal. Let's get to know these unusual—and deadly—plants.

Plants that trap, eat, and **digest** bugs are also sometimes called insectivorous plants.

# Active vs. Passive

There are over 600 known species, or kinds, of carnivorous plants. These plants differ in the way they look, how they grow, and what they eat. They even trap their **prey** differently.

Carnivorous plants have active or passive traps. Active traps are traps that move to capture their prey. One example is a trap that closes around a bug's body. Passive traps don't move to catch prey. Instead, the prey comes to the trap. This includes a trap that a bug or animal may fall into by mistake.

**Super Fact!**

Some order carnivorous plants into five different groups based on the plant's trapping method. The five trapping methods include snap, pitfall, flypaper, snare, and suction.

Carnivorous plants are found on every continent, or one of Earth's seven great landmasses, except for Antarctica!

# Clever Adaptations

Adaptations are a change in a type of plant or animal that makes it better able to live in its surroundings. The ability to eat meat—such as bugs or animals—is a clever plant adaptation.

Bugs and animals are the main food sources for carnivorous plants. This adaptation may have started long ago in plants that grew in soil that had few **nutrients**. The plants that could eat meat got more nutrients compared to those that couldn't. Over time, the adaptation was passed on to new plants.

The cobra lily uses a pitfall trap to catch prey. The plant is found in California and Oregon.

# Damp Dwellings

Most carnivorous plants live in **damp** or sandy habitats like bogs and swamps. A habitat is the natural place where an animal or plant lives. The soil in these habitats is often wet and high in **acid**. The soil is usually low in nitrogen, which is something most plants need to survive.

Most soils contain, or have, nitrogen in the form of nitrates. Plants take in nitrates through their roots. If nitrates are in low supply, plants have to find the nitrogen they need to survive elsewhere. Bugs and animals are a great source of nitrogen.

**Super Fact!**

A plant's process of making food with sunlight, water, chlorophyll, and **carbon dioxide** is called photosynthesis. Chlorophyll is a green pigment, or a substance that gives a plant or animal color.

The dewy pine is one of the only carnivorous plants that doesn't live in a wet **dwelling** or area. The plant is found only on dry hills in parts of Portugal, Spain, and Morocco.

# Plant Profile

Carnivorous plants have a long list of bugs and animals they eat. Most of these plants eat small bugs, like flies, bees, moths, and ants. But some meat-eating plants go after larger prey, including frogs and—in some uncommon cases—**rodents**!

In general, meat-eating plants are small. Species can range from a few inches to more than 3 feet (1 m) in height. Some carnivorous plants are so small that they are hard to spot among other plants in their habitat.

Sundews, butterworts, and pitcher plants are all common meat-eating plants.

sundew

pitcher plant

butterwort

# The "Snap" Trap!

Known for its spiky "teeth" that catch prey, the Venus flytrap is one of the best known carnivorous plants. The plant's trap looks like a mouth, but it is actually a set of specialized leaves.

When a bug lands on a Venus flytrap, it brushes against tiny hairs. These thin hairs prompt, or cause, the plant's trap to snap shut. The plant's leaves then give off a special liquid that breaks down the bug's body. It takes about three to five days for a Venus flytrap to digest a bug.

**Super Fact!**

Scientists say Venus flytraps can count! The plant's trap only closes if its hairs are brushed twice within a 30-second period. Also, a Venus flytrap only digests its prey if its hairs are brushed five times.

It takes less than a second for a Venus flytrap to snap shut and feast on its next meal!
SNAP!

# The Pitcher Pitfall

While the Venus flytrap actively traps bugs, other carnivorous plants take a more passive approach. Pitcher plants use their leaves to form a pitfall trap to catch their prey. They have leaves shaped like a pitcher.

The plant's pitfall trap holds nectar that draws in bugs. Drawn to the sweet smell, the prey falls inside the plant. Smooth sides or downward-pointing hairs keep the prey from crawling back out. A pool of **enzymes** lies at the bottom of the trap. The enzymes help the plant begin digesting its food.

**Super Fact!**

Some pitcher plants are endangered, or in danger of dying out. This includes the Attenborough's pitcher plant, a large carnivorous plant that grows to almost 5 feet (1.5 m) tall.

Pitcher plants have leaves in many shapes and sizes. Some are shaped like trumpets, jugs, or bowls.

# Bloodthirsty Bladderworts

Bladderwort plants are rootless, carnivorous plants that often grow underwater. These plants are found in both wet soil and bodies of water, such as ponds, lakes, and streams.

Bladderwort plants use suction "bladder" traps. Each trap has a door-like part that opens to the inside of the plant. The door gives off a sweet smell to draw in prey. If an animal gets too close, it rubs the tiny hairs surrounding the door. Suddenly—*snap*! The door flies open, and a quick flow of water sucks the animal inside.

Bladderworts are very common and live all over the world.

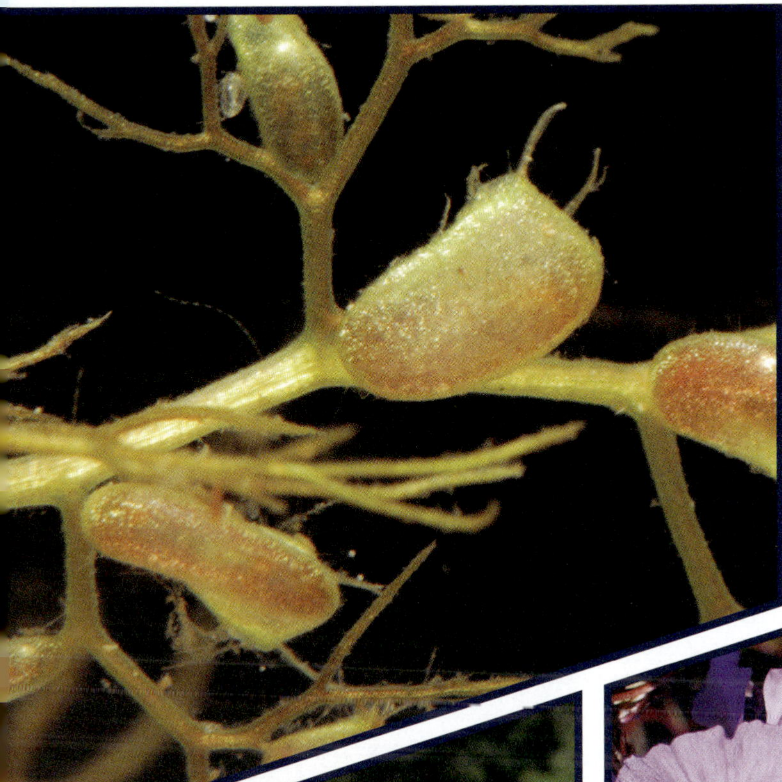

There are over 200 species of bladderworts across the world. Many are found in North America, including the common bladderwort and purple-flowered bladderwort.

# Sunny Surprise Attack!

The sundew may have a sunny name, but it's also a deadly carnivorous plant. These plants are named for small drops of nectar on their leaves that flash and shine in the sun. This liquid is at the tip of tiny, sticky hairs that cover sundew leaves. When an insect lands on the plant, it gets stuck in the plant's "flypaper" trap!

All meat-eating plants, like sundews, have **sneaky** ways to catch bugs and animals. From far away, carnivorous plants may look harmless—but bugs better watch out!

**Super Fact!**

Once a sundew catches a bug for its next meal, the plant wraps **tentacles** around the bug's body to kill it.

It only takes a sundew about 15 minutes to kill a bug to eat, but these plants might spend a few weeks digesting their food.

# GLOSSARY

**acid:** A liquid that breaks down matter.

**carbon dioxide:** A gas breathed out by animals.

**damp:** Somewhat wet.

**digest:** To break down food inside the body so that the body can use it.

**dwelling:** The place where someone or something lives.

**enzyme:** Matter made in the body that helps certain actions necessary for life to occur.

**instinct:** An inborn reaction or behavior that aids in survival.

**nutrient:** Something a living thing needs to grow and stay alive.

**prey:** An animal that is hunted by other animals for food.

**rodent:** A small, furry animal with large front teeth, such as a mouse or rat.

**sneaky:** Doing something in a quiet and secret way.

**tentacle:** A long, thin body part that sticks out from a plant or animal and is used for grabbing things.

# FOR MORE INFORMATION

## BOOKS

Braun, Eric. *Corpse Flower vs. Venus Flytrap.* Mankato, MN: Black Rabbit Books, 2018.

Hofer, Charles C. *Silent But Deadly Plants.* North Mankato, MN: Capstone Press, 2022.

## WEBSITES

**Awesome 8 Carnivorous Plants**
*kids.nationalgeographic.com/awesome-8/article/carnivorous-plants*
Discover fun facts about plants that eat meat at this website.

**How Does a Venus Flytrap Work?**
*wonderopolis.org/wonder/How-Does-a-Venus-Flytrap-Work*
Learn more about the most famous carnivorous plant here.

# INDEX